Contents

Module I

    UAS STEM English Terminology and Acronyms

        Aviation Terms

        Navigation Terms

Module II

    Math Navigation and Geography

        Units of Measure

        Reference Points

        Chart Anatomy

Module III

    Sectional Chart (Latitude - Longitude)– Model Build Lab

        Build Tools and Materials

        Latitude & Longitude Design guidelines

Module IV

    Airspace Classifications Lab

    Navigational Terms

    Airspace Classification Descriptions

Module V

    Principals of Model Construction

    Understanding the Basics of Structural Design

Module VI

    Airspace Model Build

    Model Build Guidelines

    Parts Inventory

Introduction

Welcome to S.T.E.A.M. LABS National Airspace Systems Uncrewed Aviation Science course. This course is presented in 6 modules In this course, you will gain the basic knowledge which is required to navigate on the planet Earth .

You will also have hands on practical experience s with basic structural engineering design and construction while exploring the science of balance, gravity and ballast. The knowledge which you gain from these topics will be combined with the science of weather and Aerodynamics .

Finally you will learn about the use of basic software applications to assist in your navigation flight skills and design proof of concept methods

Once you have completed this S.T.E.A.M. LABS course you be ready for one of our other science, technology, engineering , aviation or math courses.

Course and Lab Objectives

I) Identification and utilization of all available resource materials and tools.

II) Application of learned core math, geography, reading comprehension and basic science skills.

III) Development of effective communication and time management skills.

IV) Present and explain navigation lab build model functionality and relationship to distance and area.

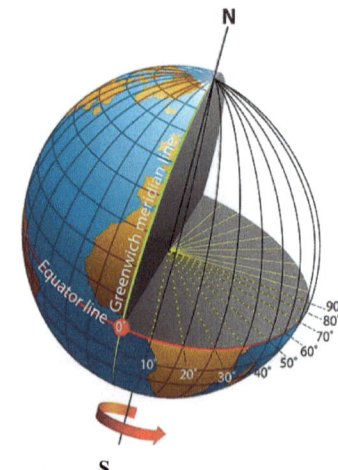

First print edition 1/10/24

Author: Michael & Mykalah Johnson

Module 1

Navigation Terms

TERM / WORD	DEFINITION
Antidotal	Something that will take away or reduce the bad effects of something unpleasant
Applicable	Affecting or relating to a person or thing
Azimuth	An azimuth is the direction measured in degrees clockwise from north on an azimuth circle.
Hemisphere	One of two halves of the earth, especially above or below the equator
Magnetic	Possessing an extraordinary power or ability to attract.
Precludes	To prevent something or make it impossible, or prevent someone from doing something
Scenario	A description of possible actions or events in the future
Criteria	Rules, Examples, Standards
Tier	One of several layers or levels
Latitude	The distance measured north or south from the Equator to the North or South Poles of the Earth. Latitude is measured in degrees, minutes and seconds.
Longitude	The angular distance east or west of the prime meridian that stretches from the North Pole to the South Pole and passes through Greenwich, England. Longitude is measured in degrees, minutes, and seconds.
Statute Mile	A Statute Mile is measured on land. It is 5,280 feet long.
Nautical Mile	A Nautical Mile is measured on the sea and in the air. It is 6,076 ft in length.

Module 1

Aviation Terms

Assorted Definitions used in Drone Aviation

Terminology	Definition
Small Uncrewed Aircraft	An Uncrewed aircraft weighing less than 55 pounds, including everything that is onboard or otherwise attached to the aircraft, and can be flown without the possibility of direct human intervention from within or on the aircraft
Small Uncrewed Aircraft System (sUAS):	This is a small UA and its associated elements
Uncrewed Aircraft (UA)	An aircraft operated without the possibility of direct human intervention from within or on the aircraft
Visual Observer (VO)	A person acting as a flight crew member who assists the small UA remote PIC and the person manipulating the controls to see and avoid other air traffic or objects aloft or on the ground
Control Station (CS)	This is your remote control for your aircraft
Model Aircraft	This is a fixed wing or helicopter Uncrewed aircraft that is: Capable of sustained flight in the atmosphere; flown within visual line-of-sight (VLOS) of the person operating the aircraft flown for hobby or recreational purposes
Terminal Radar Service Area (TRSA)	This is the 5 mile radius area that is covered by radar surrounding airports with an Air Traffic Control Tower that provides takeoff and landing instructions for airplanes. There are 3 different classes of airports with radar: Class B, Class C, and Class D
Remote Pilot in Command Remote PIC or Remote Pilot	This is a person who holds a remote pilot certificate with an sUAS rating and has the final authority and responsibility for the operation and safety of an sUAS operation conducted under part 107

Industry Acronyms & Terminology Aviation

Abb./Acronym	Definition
AGL	Above Ground Level
ATC	Air traffic Control
FAA	Federal Aviation Administration
FL	Flight Level
MSL	Mean Sea Level
NASA	National Aeronautical Space Administration
NAS	National Airspace System
NM	Nautical Miles
PIC	Pilot in Command
SM	Statue Miles
TN	True North
TRSA	Terminal Radar Service Area
UAV	Un-Crewed Aircraft Vehicle
VO	Visual Observer

Module I Quiz

1. What is the definition of AGL ?

   A) Altitude Gain Level
   B) Aircraft Glide Line
   C) Above Ground Level

2. What is the definition of MSL ?

   A) Mean Sea Level
   B) Main Sea Level
   C) Moving Sea Level

3. What is the definition of Azimuth ?

   A) An azimuth is measured in degrees counter clockwise from north on an azimuth circle
   B) An azimuth is the direction measured in inches clockwise from south on an azimuth circle
   C) An azimuth is the direction measured in degrees clockwise from north on an azimuth circle

4. What is the definition of Hemisphere ?

   A) One of two halves of the earth, especially above or below the equator
   B) One of two halves of the earth, especially east or west of the Prime Meridian
   C) Both A & B
   D) None of the above

5. What is the distance of a Statute Mile ?

   A) 6,076 feet
   B) 5,280 feet
   C) 2,640 feet

6. What is the distance of a Nautical Mile ?

   A) 6,076 feet
   B) 3,780 feet
   C) 7, 250 feet

NOTES

Module II

Math Navigation and Geography

1) Surface Distance is measured by using:

    a) Tape measurer. The tape measurer can be broken down from feet to inches to ¼ of an inch or as little as 1/16th of an inch. As shown below.

    b) Number Line. A number can be used to measure time or space since it is infinite. The units of measurement can range from less than zero (nano) to greater than One Trillion (Tetra).

PREFIX	SYMBOL	MULTIPLIER	EXPONENT FORM
exa	E	1,000,000,000,000,000,000	10^{18}
peta	P	1,000,000,000,000,000	10^{15}
tera	T	1,000,000,000,000	10^{12}
giga	G	1,000,000,000	10^{9}
mega	M	1,000,000	10^{6}
kilo	k	1,000	10^{3}
hecto	h	100	10^{2}
deca	da	10	10^{1}
Basic Unit	Basic Unit	1	10^{0}
deci	d	0.1	10^{-1}
centi	c	0.01	10^{-2}
milli	m	0.001	10^{-3}
micro	μ	0.000,001	10^{-6}
nano	n	0.000,000,001	10^{-9}
pico	p	0.000,000,000,001	10^{-12}
femto	f	0.000,000,000,000,001	10^{-15}
atto	a	0.000,000,000,000,000,001	10^{-18}

Math Navigation and Geography

Units of Measure

2) The shape of the earth is used in Aviation for measuring distance by using:

a) Lines of longitude which measures the distance around the earth from east to west. The lines of longitude connect the North Pole and South Pole. These lines of longitude are called Meridians. The zero line of longitude is called the Prime Meridian and it runs through a town in England called Greenwhich, England.

Lines of latitude run just like a number line from the number 0, which is the Prime Meridian west to 180 and east 180. The latitude number lines are measured in degrees from number to number. Just like a number line there are smaller units of measurement between degree lines of latitude which are broken down into minutes and seconds.

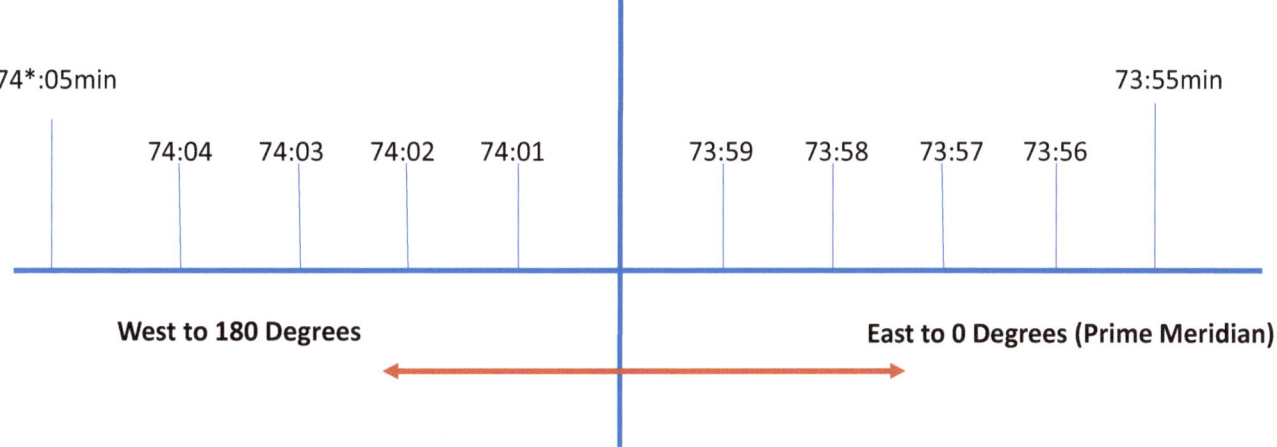

Latitude degree distance:

Each degree of latitude is approximately **69 miles (111 kilometers)** apart. At the equator, the distance is 68.703 miles (110.567 kilometers). At the Tropic of Cancer and Tropic of Capricorn (23.5 degrees north and south), the distance is 68.94 miles (110.948 kilometers).

Longitude degree distance:

At the equator, longitude lines are the same distance apart as latitude lines — one degree covers about **111 kilometers (69 miles)**. But, by 60 degrees north or south, that distance is down to 56 kilometers (35 miles). By 90 degrees north or south (at the poles), it reaches zero.

$1/60^{th}$ of a degree is approximately one (nautical) mile.

Math Navigation and Geography

Units of Measure

2) The shape of the earth is used in Aviation for measuring distance by using:

    b) Lines of Latitude which measure the distance from the Equator 90 degrees north and south to the North and South Poles in the eastern and western hemispheres of the earth. The Equator is the zero line of latitude which separates northern and Southern hemispheres. Lines of Latitude circle the earth and never meet. They get smaller as you approach the North and South Poles. The latitude number lines are measured in degrees from number to number. Just like a number line there are smaller units of measurement between degree lines of latitude which are broken down into minutes and seconds

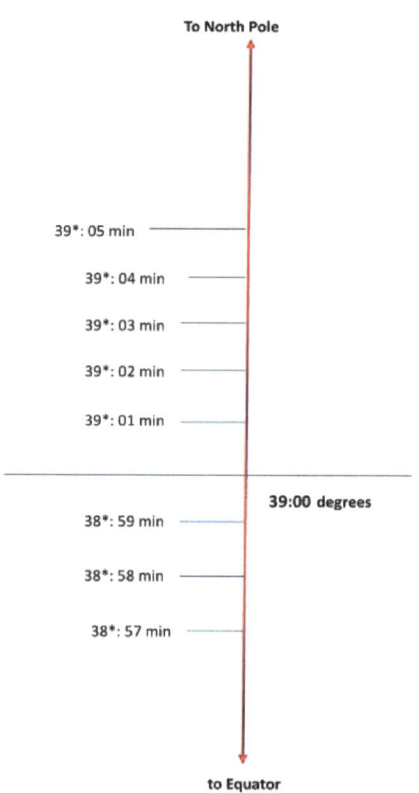

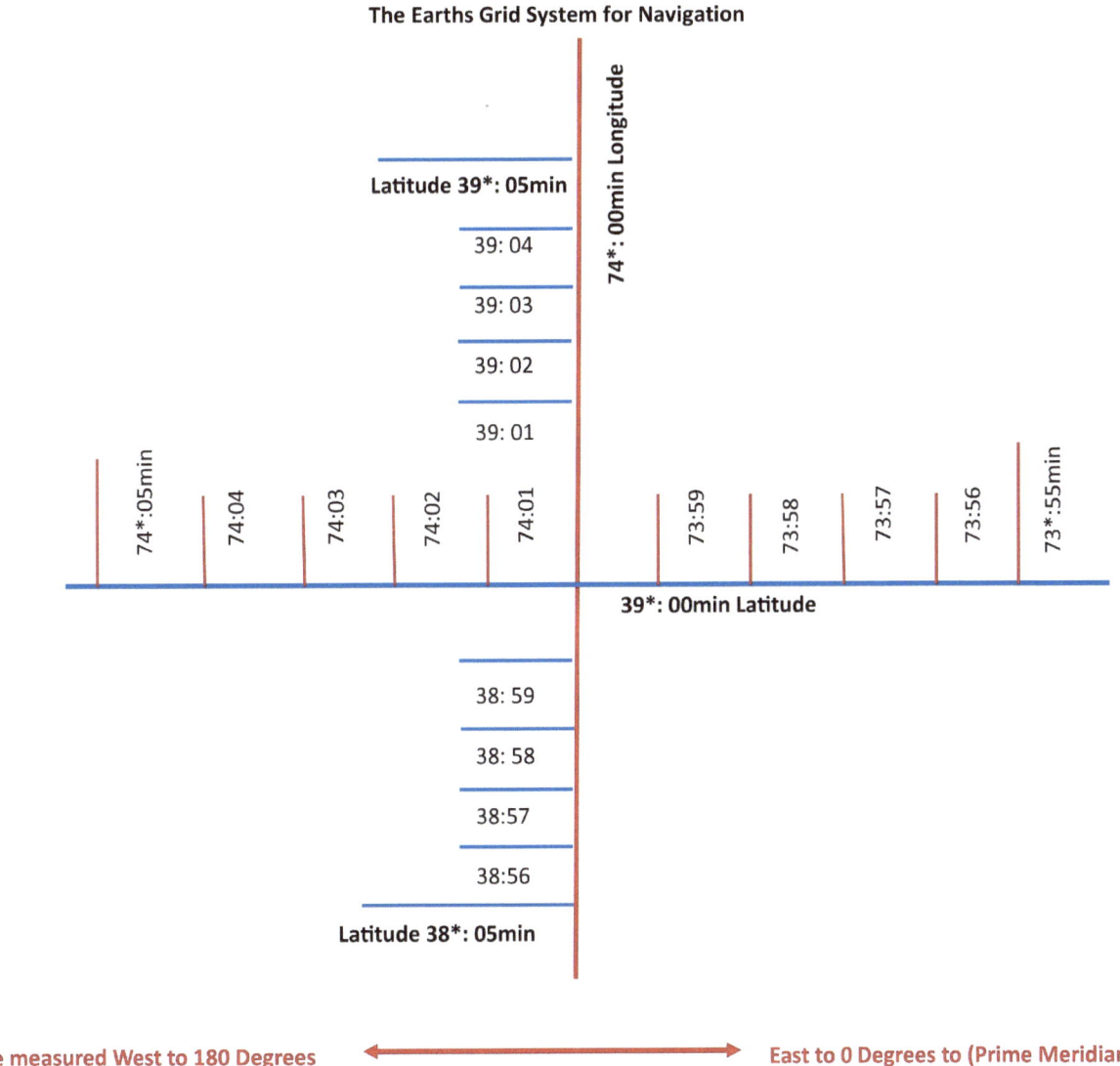

Module II
Section B
Navigational Chart Anatomy

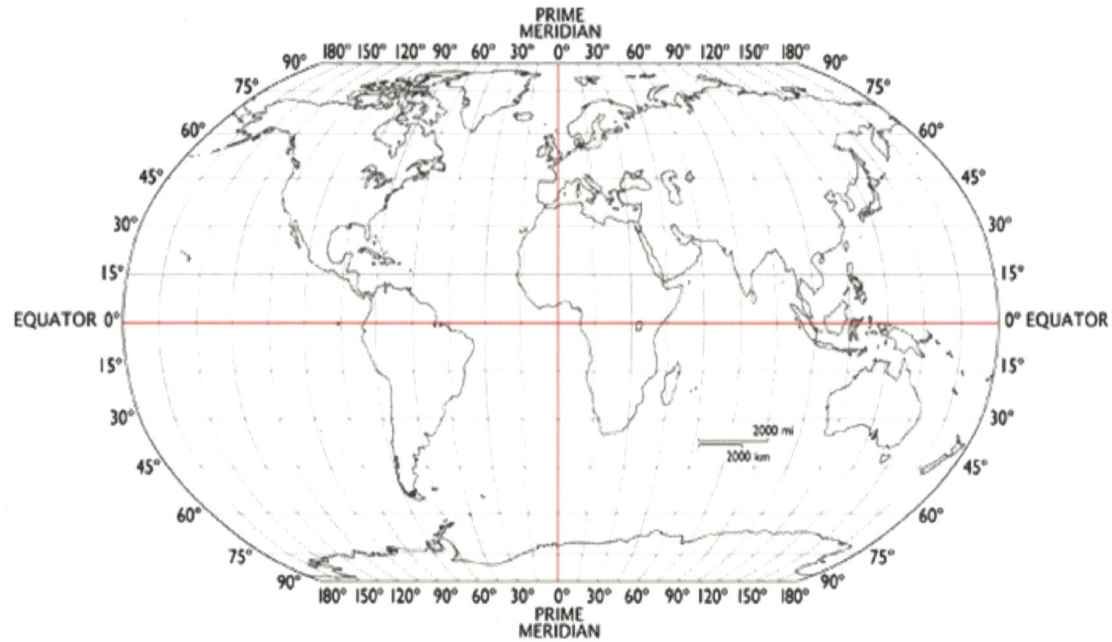

The Equator circles the entire planet and **is the zero-degree line of latitude**. It is equal distance from the North and South Poles. It runs through two continents, Africa and South America. It also runs through three island nations in the Pacific Ocean. **The Equator intersects the Prime Meridian in the Atlantic Ocean** off the west coast of Africa.

The Prime Meridian connects the North Pole to the South Pole, it **is the zero degree of longitude**, it separates the Western Hemisphere from the Eastern Hemisphere.

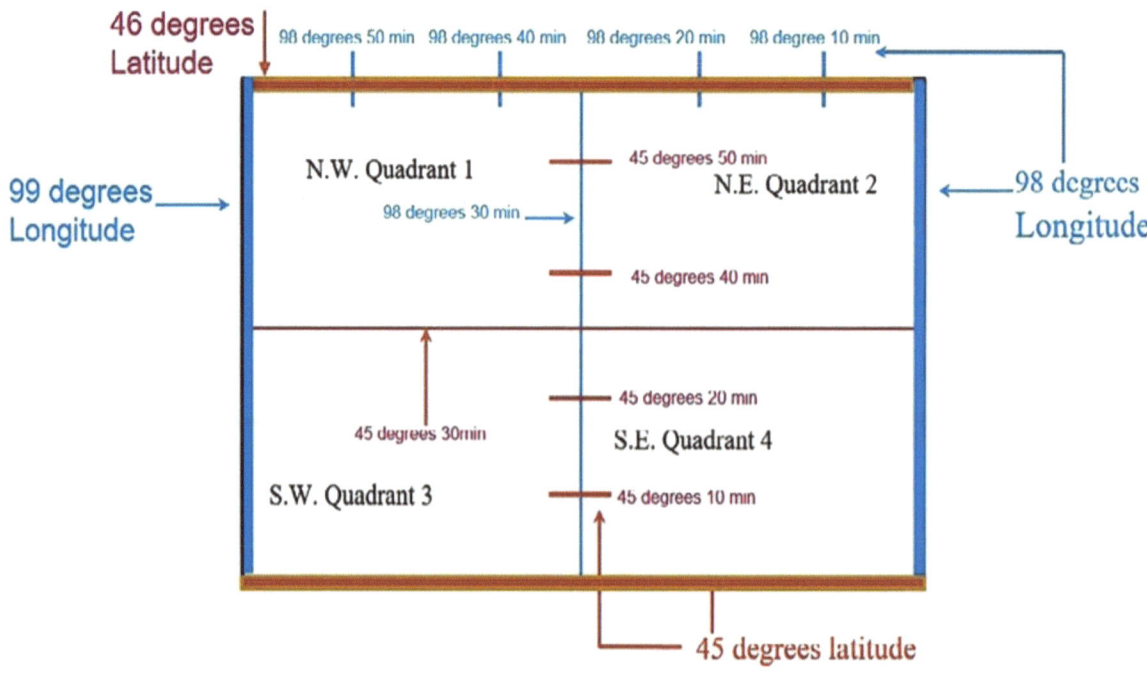

NAS-UAS Introduction

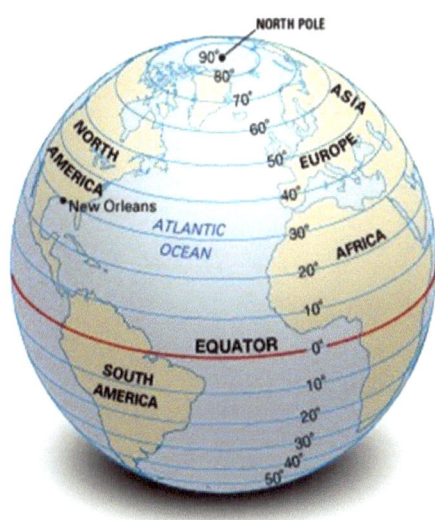

FACTS ABOUT LINES OF LATITUDE

- Are known as parallels.
- Run in an east-west direction.
- Measure distance north or south from the Equator.
- Are parallel to one another and never meet.
- Cross the prime meridian at right angles.
- Lie in planes that cross the Earth's axis at right angles.
- Get shorter toward the poles, with only the Equator, the longest, a great circle.

FACTS ABOUT LINES OF LONGITUDE

- Are known as meridians.
- Run in a north-south direction.
- Measure distance east or west of the prime meridian.
- Are farthest apart at the Equator and meet at the poles.
- Cross the Equator at right angles.
- Lie in planes that pass through the Earth's axis.
- Are equal in length.
- Are halves of great circles.

THE EARTH'S GRID SYSTEM

- Only the city of New Orleans, La., is located at the crossing of the 30th east-west line north of the Equator and the 90th north-south line west of the prime meridian.
- Lines of latitude cross lines of longitude at right angles.
- Although only a few lines of latitude and longitude are shown on globes and maps, their number is infinite.

Module II Quiz

1. There are 60 minutes from one degree of latitude to the next degree of latitude, each minute of each degree is equal to approximately how many nautical miles ?

    A) 1 mile

    B) 30 miles

    C) 10 miles

2. If a tape measurer or ruler is divided into 16 parts (1/16 –16/16) from inch to inch , how many 1/16th of one inch would it take to make 1/4 of an inch ?

    A) 5/16

    B) 6/16

    C) 4/16

3. Lines of Latitude measure the distance from which starting point ?

    A) The Prime Meridian

    B) The Equator

    C) The Tropic of Cancer

4. Lines of Longitude measure the distance from which starting Point ?

    A) The Tropic of Capricorn

    B) The Tropic of Cancer

    C) The Prime Meridian

5. Lines of Latitude measure distances in which direction ?

    A) North and South from the Equator

    B) East and West of the Equator

    C) East and West of the Prime Meridian

6. Lines of Longitude measure distance in which direction ?

    A) North and South from the Prime Meridian

    B) East and West from the Prime Meridian

    C) North and South of Tropic of Cancer

Module III

Latitude – Longitude Surface Grid Model Build Lab

Overview

The goal of this Model build lab is to give you a hands-on view and understanding of how the degrees of latitude and longitude work together to build maps for finding your way from one location to a different location while flying through the air.

Model Build Rules

1. You must find out the latitude and longitude degrees for the location you are building the grid model at.

2. Your grid model must have (4) degree lines of Latitude and (4) degree lines of longitude.

3. When building your you must identify which lines are latitude and which lines are longitude and number them in proper direction, in order using the equator as zero line of Latitude and the Prime Meridian as zero line of longitude.

4. You must include ground point connectors for future airport airspace classes

5. Your PVC Piping building materials will be listed on numbered chart in the order they must be used.

6. Use the supplied drawing and Building materials parts list to complete the model build.

Module III
Navigation Lab – Model Build

Tool Description	Used For:
25 foot retractable Tape Measurer	Measuring PVC pipe
48" Paper Tape Measurer	Measuring Pex Tubing
Pressure Grip Pliers	Removing PVC Fittings
Pipe Level (clamp on type)	Verify level connections
PVC Pipe Cutter	Trimming PVC Pipes
Culinary Scissors	Cutting Pex Tubing
Fine Sand Paper	Smoothing cut edges of PVC Pipie

Latitude – Longitude Surface Grid Model Building Materials

Location Number	PVC Pipe Size	Right Side Connector	Left Side Connector
1	23" Length	3P1V	H90T
2	11" Length	H90T	4P1V
3	11" Length	4P1V	4P1V
4	23" Length	4P1V	3P1V
5	23" Length	V90T	3P1V
6	23" Length	4PH	V90T
7	11" Length	V90T	4P1V
8	11" Length	5P1V	4P1V
9	23" Length	H90T	3P1V
10	11" Length	V90T	5P1V
11	11" Length	5P1V	3P1V
12	11" Length	4P1V	V90T
13	11" Length	5P1V	5P1V
14	11" Length	4P1V	3P1V
15	23" Length	H90T	4PH
16	11" Length	4PH	4P1V
17	11" Length	4P1V	5P1V
18	11" Length	5P1V	4P1V

Latitude – Longitude Surface Grid Model Building Materials

Location Number	PVC Pipe Size	Right Side Connector	Left Side Connector
19	11" Length	4P1V	V90T
20	11" Length	V90T	H90T
21	11" Length	V90T	4PH
22	23" Length	5P1V	5P1V
23	11" Length	H90T	H90T
24	11" Length	H90T	V90T
25	11" Length	5P1V	V90T
26	11" Length	H90T	V90T
27	23" Length	5P1V	H90T
28	11" Length	5P1V	V90T
29	11" Length	V90T	5P1V
30	23" Length	5P1V	H90T
31	23" Length	3P1V	H90T
32	23" Length	H90T	5P1V
33	23" Length	H90T	5P1V
34	23" Length	3P1V	H90T
35	23" Length	3P1V	H90T
36	11" Length	H90T	V90T

Latitude – Longitude Surface Grid Model Building Materials

Location Number	PVC Pipe Size	Right Side Connector	Left Side Connector
37	11" Length	V90T	H90T
38	23" Length	H90T	3P1V

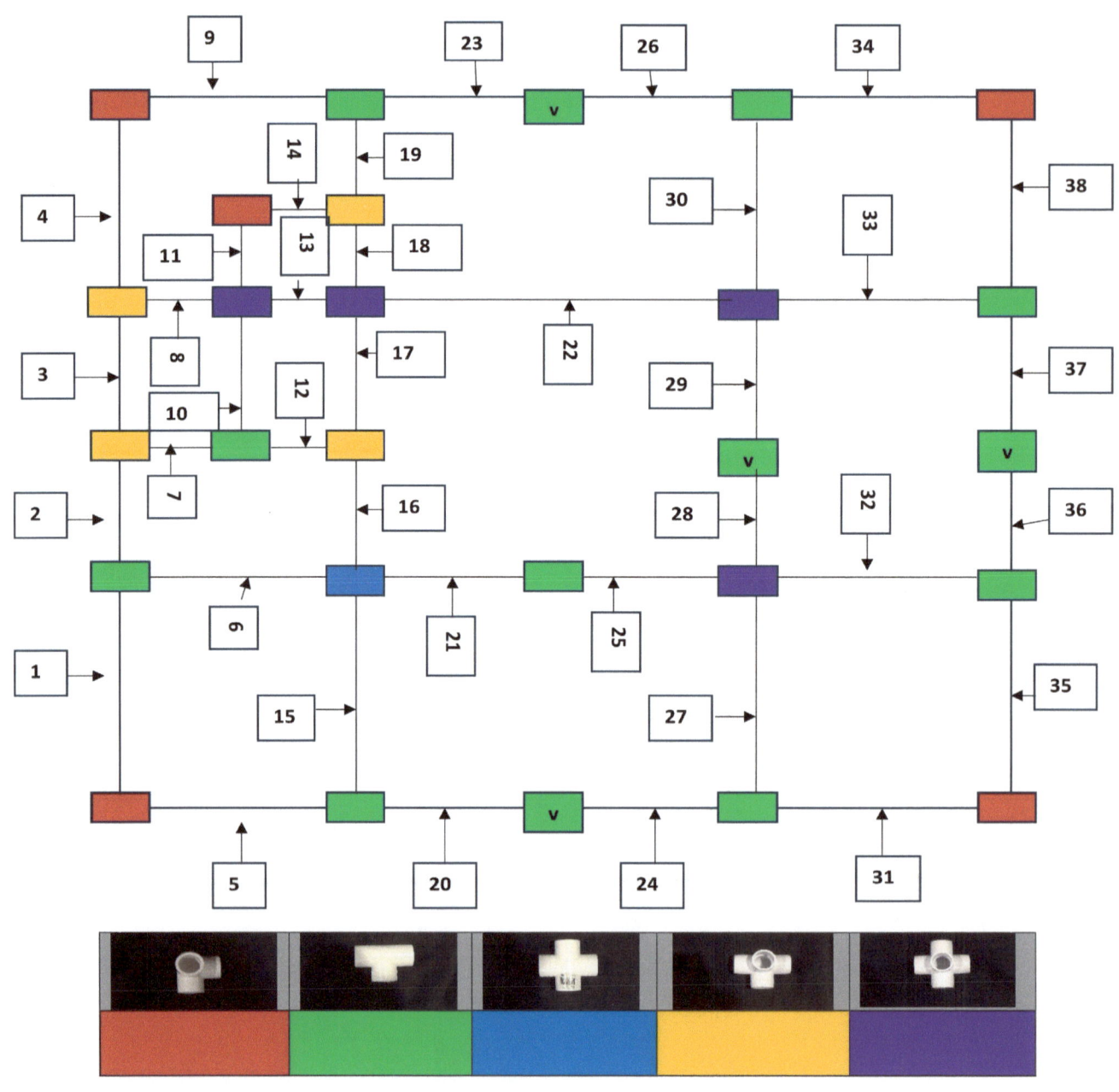

SURFACE LAT-LONG BASE DIAGRAM

NOTES

Latitude-Longitude Model Perimeter Inventory List
Ground Level Latitude & Longitude Surface Grid

GROUND LEVEL PART DESCRIPTION	SIZE	QUANTITY
1-4 Perimeter PVC Pipe Base Wall	½" Diameter X 23" Length ½" Diameter X 11" Length	2 2
5-31 Perimeter PVC Pipe Base Wall	½" Diameter X 23" Length ½" Diameter X 11" Length	2 2
9-34 Perimeter PVC Pipe Base Wall	½" Diameter X 23" Length ½" Diameter X 11" Length	2 2
35-38 Perimeter PVC Pipe Base Wall	½" Diameter X 23" Length ½" Diameter X 11" Length	2 2
3P1V Corner Connector	½" Diameter	5
4P1V Connector	½" Diameter	4
4PH Connector	1/2" Diameter	1
T90 Connector	½ " Diameter	13
5P1V Connector	½" Diameter	4

AIRSPACE WALL   1-4 Build Model

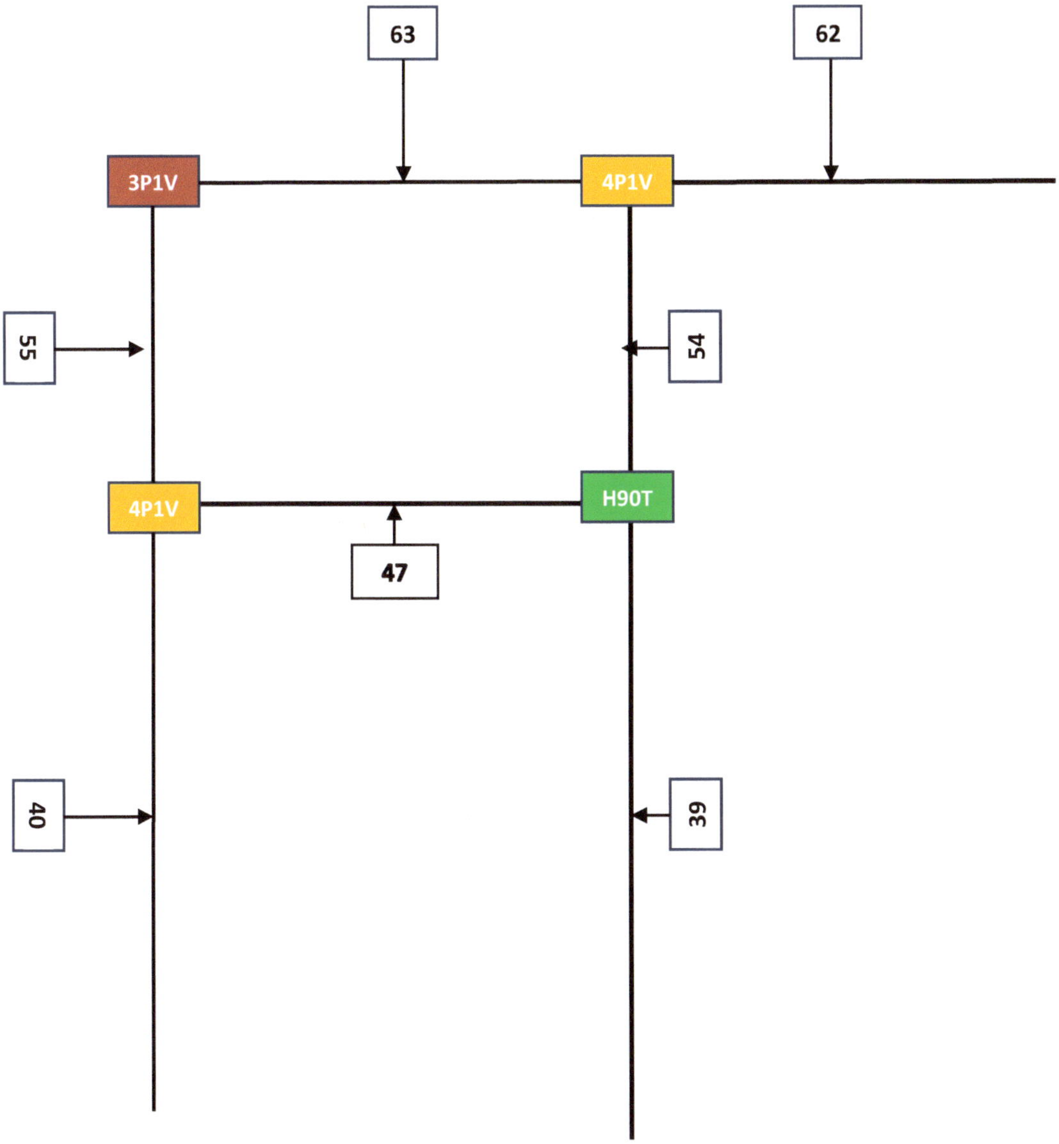

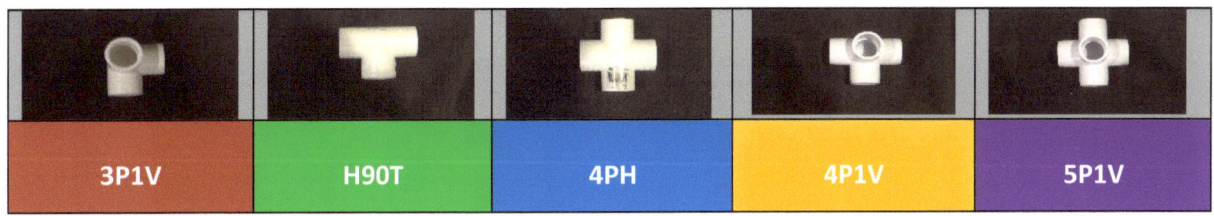

AIRSPACE WALL   9-34 Build Model

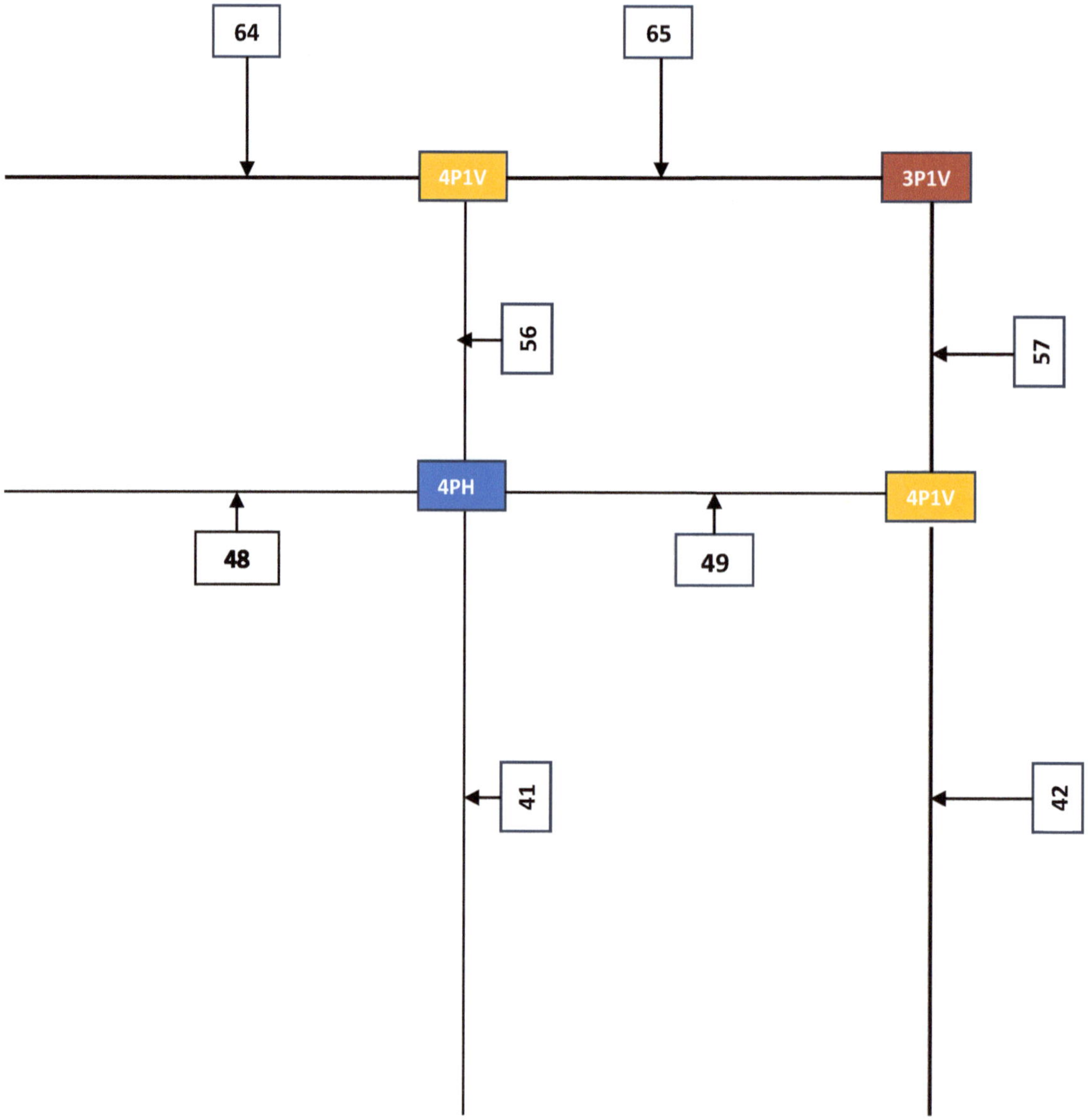

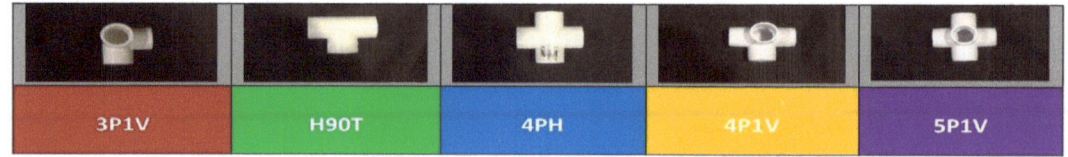

AIRSPACE WALL   5-31 Build Model

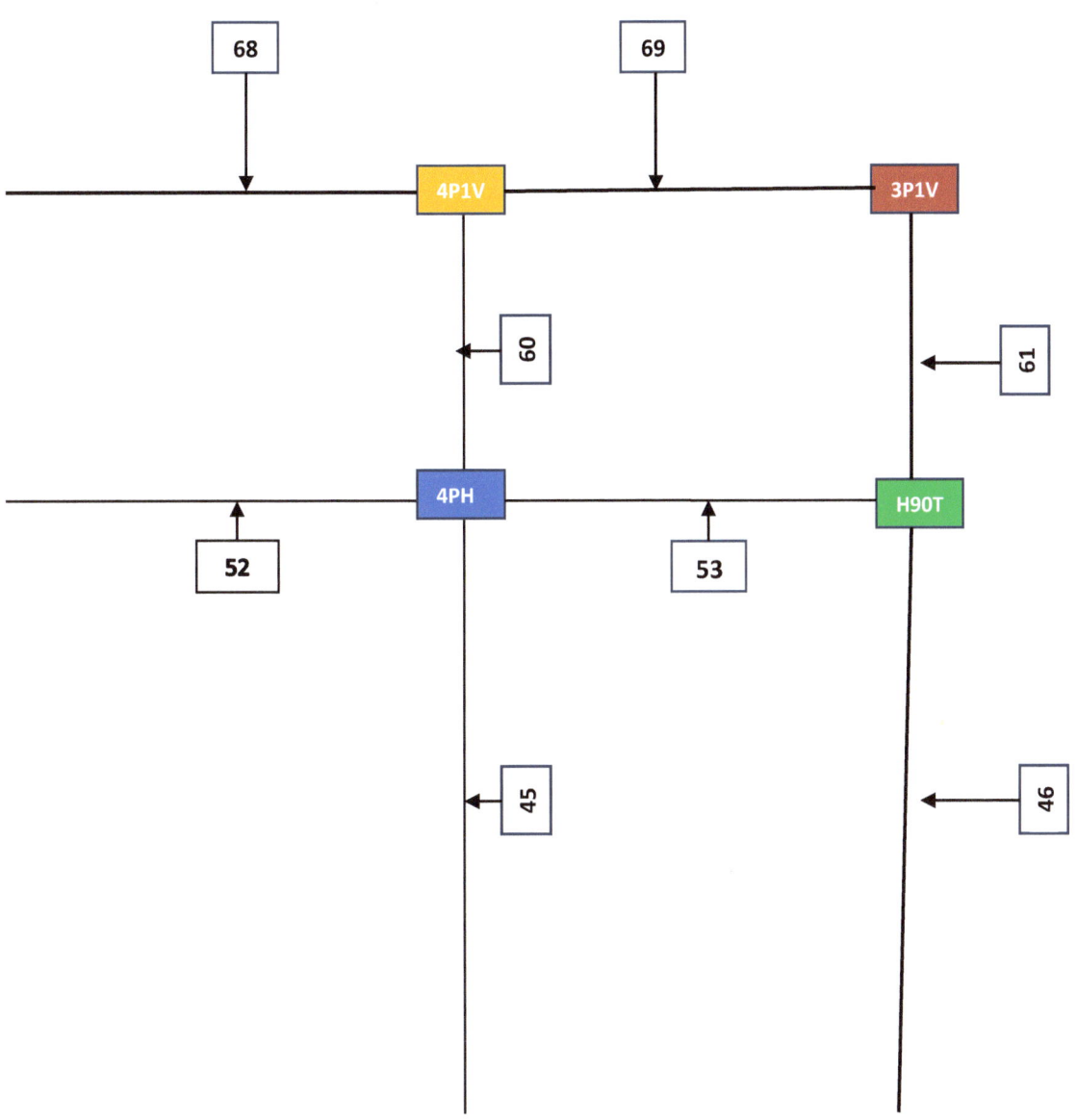

AIRSPACE WALL 35-38 Build Model

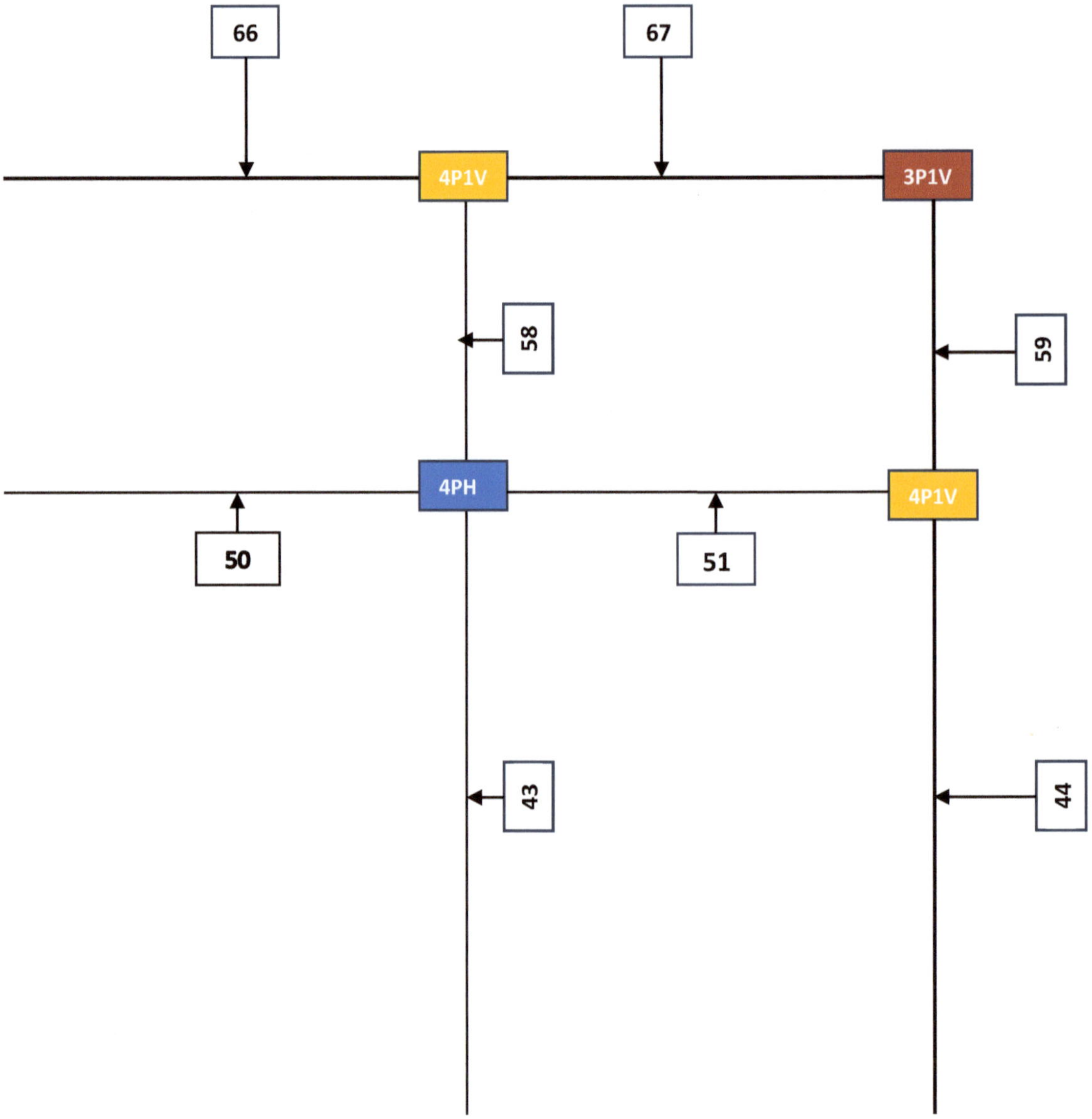

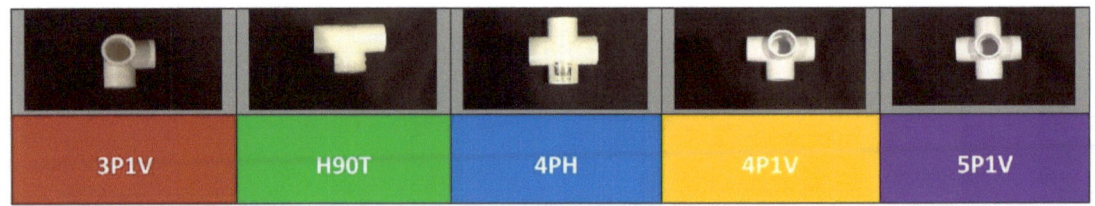

AIRSPACE CEILING 70-73 Build Model

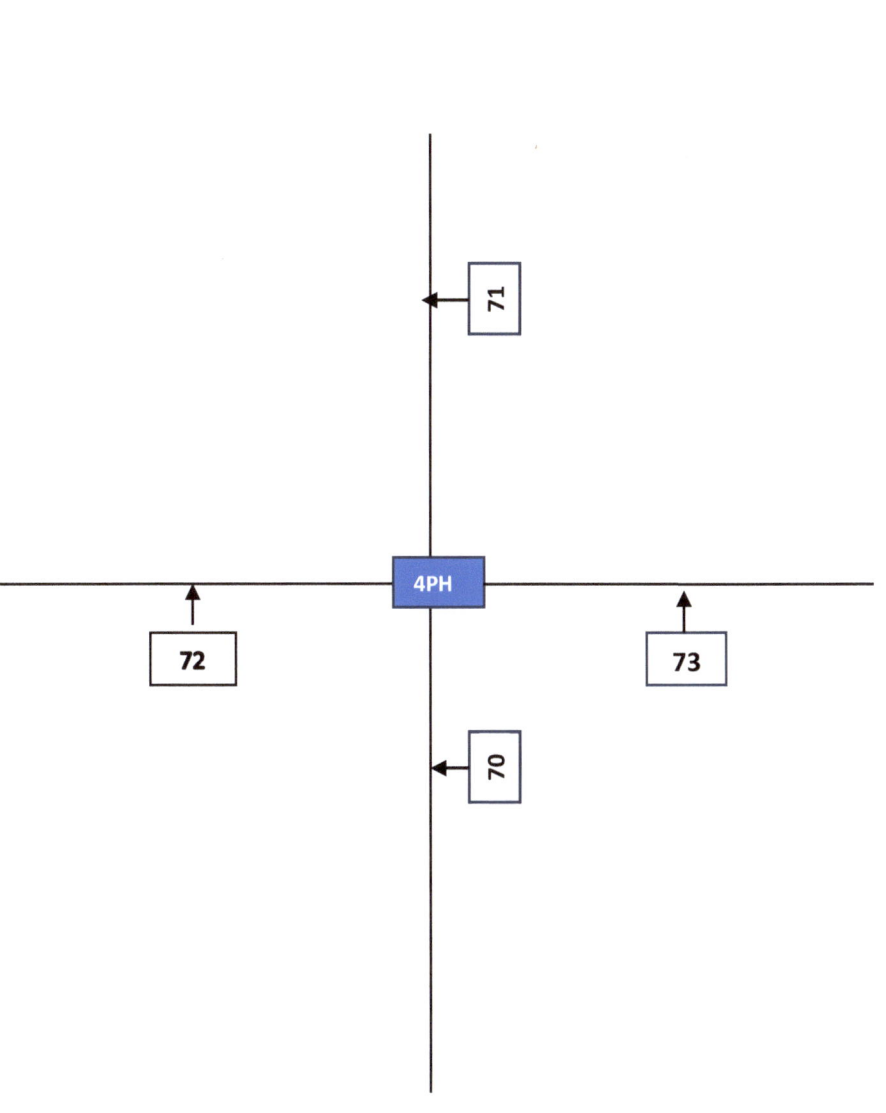
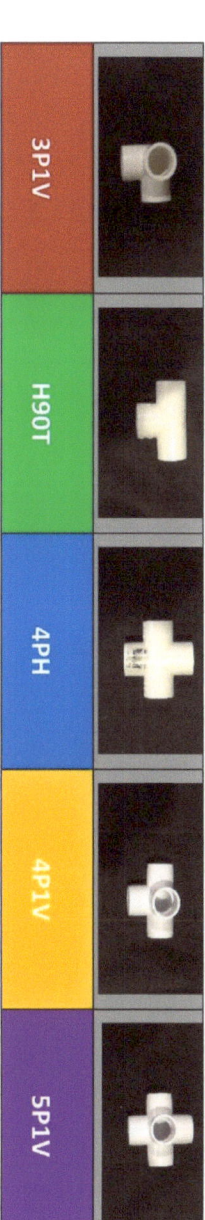

NOTES

Latitude & Longitude Model
Facts & Observations

1) What is the actual Latitude and Longitude coordinates that your model is built on?

   LAT: Degrees_____   Minutes_____   Seconds_____

   LONG: Degrees_____   Minutes_____   Seconds_____

2) What is the most western value of Longitude _____

3) What is the most southern degree of Latitude _____

4) How many nautical square miles does your model cover _____

************BONUS CHALLENGE **********

5) If a 3 bedroom house takes up 1600 square feet, how many 1600 square foot homes could you build in a 900 square mile area of 1 degree latitude by 1 degree longitude sectional chart?

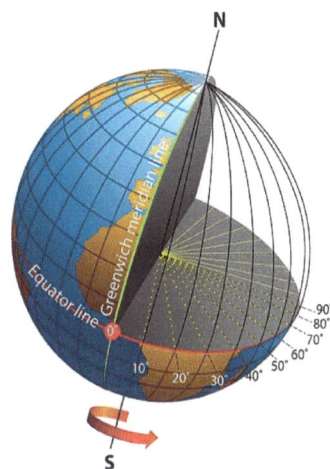

You find access to S.T.E.A.M. LABS courses on the following websites:

https://steamthrudrones.com/ https://dronelecture.com/

Our direct contact : www.steamlabsorg@gmail.com

Module IV
Airspace Classifications

INTRODUCTION

Understanding the Course and Lab Objectives

I) Identification, utilization and integration of all build materials introduced in this airspace lab with the previously constructed Surface Navigation grid model build.

II) Apply learned core math, geography, reading comprehension and basic science skills.

III) Development of effective communication and time management skills.

IV) Present and explain lab structural model functionality in an airspace classification environment.

V) Create an alternative airspace model by relocating airspace class elevations from original latitude and longitude drawing coordinates to new envisioned latitude and longitude coordinates.

TABLE of CONTENTS

Module I	Module II	Module III
Navigational Terms	**Principals of model Construction**	**Airspace Build Lab**
Airspace Classifications	**Basics of Structural Design**	**Build Tools and Materials**
		Airspace Classification Design Guidelines
A) Class B	A) Scale Conversions	
B) Class C	B) Inches- Feet	A) Airspace Model Build
C) Class D	C) Review Quiz	B) Inventory List
D) Class E		C) Challenge Questions
E) Review Quiz		

    You must read the build requirement measurements carefully. The PVC pipe kits are not cut to the exact length. This will require you to measure each pipe and cut it the correct length using the supplied retractable tape measurer and PVC pipe cutter tool. The Pex tubing which is supplied in blue and red will have to be cut using the supplied cutting tool for Pex. Airspace vertical elevation PVC pipes should be mounted diagonally from each other and not linear. This will allow for a flat landing surface for the assembled Lite-B Drone landing pad. The total Length and width of the navigation grid surface must have the same total length and width of the walls and ceiling PVC piping to ensure stability of the model.

Module IV

Navigational Terms

TERM/WORD	DEFINITION
Balance	*Balance* is having the right amount — not too much or too little
Ballast	*Aeronautics*. something heavy, to give steadiness to; keep steady:
Collaboration	Collaboration is the process of working together towards a common goal.
Elevation	The height to which something is elevated or to which it rises:
Environment	Location
Hemisphere	One or two halves of the Earth, especially above or below the equator
Magnetic	Possessing an extraordinary power or ability to attract
Precludes	To prevent something or make it impossible, or prevent someone from doing something
Scenario	A description of possible actions or events in the future
Criteria	Rules, examples, standards
Tier	One of several layers or levels
Latitude	The distance measured north or south from the Equator to the North or South Poles of the Earth. It is measured in degrees, minutes, and seconds.
Longitude	The angular distance east or west of the Prime Meridian that stretches from the North Pole to the South Pole and passes through Greenwich, England. It is measured in degrees, minutes, and seconds.
Statute Mile	It is 5,280 feet long and is measured on land
Structural	having structure or a structure, Structure: the way in which the parts of a system or object are arranged or organized
Nautical Mile	It is 6,076 feet in length and is measured on the sea and in the air
Verti-Port	a defined area that can support the landing and take-off of eVTOL aircrafts during flight operations

MODULE IV

Airspace Classifications

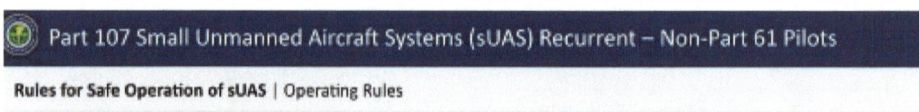

FAA rules apply to the entire National Airspace System – there is no such thing as "unregulated" airspace.

AIRSPACE CLASS DESCRIPTIONS

CLASS	DESCRIPTION
FL 600	**FL600 is the approximate top of the troposphere,** altitudes above this pose a much higher safety risk to commercial activities because it offers very little promise for survivability of humans without a full pressure suit in the event of malfunctions of aircraft systems (even if oxygen is available). Traditional aircraft engines also lose efficiency above the troposphere (surface to **approximately 60,000 feet**) and operating a vehicle at those altitudes poses significant aerodynamic issues that make commercial operations less profitable than traditional operations.
A	**This is the airspace from 18,000 feet mean sea level (MSL) up to and including flight level (FL) 600.** This includes the airspace overlying the waters within 12 nautical miles (NM) of the coast of the 48 contiguous states and Alaska
B	**Area is individually tailored, can consist of two or more layers (which makes it look like an upside-down wedding cake),** and is designed to contain all published instrument procedures once an aircraft enters into the airspace. **This is the airspace from the surface to 10,000 feet MSL surrounding the nation's busiest airports.** An ATC clearance is required for all aircraft to operate in the area. All aircraft that are cleared receive separation services within the airspace
C	**The class C airspace is individually tailored, the airspace usually consists of a surface area with five NM radius, an outer circle with a ten NM radius that extends from 1,200 feet to 4,000 feet above the airport elevation, and an outer area.** This is the airspace from the surface to 4,000 feet above the airport elevation (charted in MSL) surrounding those airports that have an operational control tower. Are serviced by a radar approach control, and have a certain number of IFR operations or passenger enplanements. Pilots must establish two-way radio communications with the ATC facility before entering the airspace and continue those communications while in the airspace

D		This is the airspace from the surface to 2,500 feet above elevation (charted in MSL) surrounding those airports that have an operational control tower. Arrival extensions for instrument approach procedures (IAPs) may be for Class D or Class E airspace. Unless otherwise advised, pilots must keep two-way radio communications with the ATC facility, and must continue those communications while in that airspace
E		This airspace extends upwards from either the surface or a designated altitude to the overlying or adjacent controlled airspace. Class E also contains: 1. **Federal airways airspace Beginning at either 700 or 1,200 feet above ground level (AGL)** used to transition to and from the terminal or en route environment 2. En route domestic and offshore airspace designated below 18,000 feet MSL Unless designated at a lower altitude, **Class E airspace begins at 14,500 MSL over the United States,** including that airspace overlying the waters within 12 NM of the coast of the 48 contiguous states and Alaska, up to but not including 18,000 feet MSL, and the airspace above FL 600
G		Class G airspace is the portion of airspace that has not been classified as Class A, B, C, D, or E, because of this, it is considered uncontrolled airspace. **This airspace extends from the surface to the base of the overlying Class E airspace** Even though there is no ATC monitoring this airspace, pilots should remember that there are visualflight rules (VFR) minimums which still apply

AIRSPACE CLASSICATIONS WHICH HAVE AIRPORTS

CLASS	DESCRIPTION
B Air Traffic Control Tower	Class B airports have three levels of airspace. The lowest level starts at the ground and goes up 1,200 feet above the airport and extends out five miles in all directions from the airport in a circle. The second level starts at 1,200 feet above the airport then extends up to 4,000 feet above the airport and extends out 10 miles in all directions from the airport. The third/top level starts at 4,000 feet above the airport then extends up to 10,000 feet and extends out 20 miles in all directions from the airport.
C Air Traffic Control Tower	Class C airports have two levels of airspace. The lowest level starts at the ground and goes up 1,200 feet above the airport and extends out five miles in all directions from the airport in a circle. The second level starts at 1,200 feet above the airport then extends up to 4,000 feet above the airport and extends out 10 miles in all directions from the airport.
D Air Traffic Control Tower	Class D airports have one Level which starts at the ground and goes up to 2,500 feet above the airport and extends out 5 miles from the airport in all directions
E No ATC Tower	Airports with no air traffic control tower have either private or public use runways and are located in areas where Class E airspace starts at 700 feet above the ground at the airport and extends out 5 miles in all directions.

MODULE IV QUIZ

1. Which airspace classifications have airports?

    a) Class A, B, and C
    b) Class B, C, and D
    c) Class B, C, D, and E

2. Which airports have air traffic control towers?

    a) Class B, C, and D
    b) Class C, D, and E
    c) Class D, E, and G

3. How high from the ground does Class B airspace rise?
    a) 2,500 ft MSL
    b) 10,000 feet MSL
    c) 4,000 feet MSL

4. How high from the ground does Class C airspace rise?
    a) 2,500 ft MSL
    b) 10,000 feet MSL
    c) 4,000 feet MSL

5. How high from the ground does Class D airspace rise?
    a) 2,500 ft MSL
    b) 10,000 feet MSL
    b) 4,000 feet MSL

MODULE V

Understanding the LAB Model (POC) Principles of Construction

The basic principles of constructing an airspace classification model are:

Safety: The main goal of an airspace classification model is to ensure the safety of all aircraft, both in the air and on the ground. This means that the model should take into account factors like the altitude of the aircraft, the type of aircraft, and the type of airspace the aircraft is in.

Efficiency: The model should also be designed to be efficient, so that it is easy for pilots and air traffic controllers to understand and follow. This means that the different categories of airspace should be clearly defined, with clear rules and regulations for each one.

Flexibility: The model should also be flexible, so that it can adapt to changing circumstances and conditions. For example, if there is a natural disaster or other emergency, the model should be able to accommodate changes in the airspace to ensure the safety of all aircraft.

Collaboration: The model should also be designed to facilitate collaboration between different agencies and organizations that are responsible for managing the airspace. This includes air traffic controllers, pilots, and other aviation professionals.

So, to summarize, an airspace classification model is a way of dividing up the air around us into different areas based on how it's used. We can make a model of this using PVC pipes and PEX tubing, with different colors representing the different classes of airspace.

Understanding the Basics of Structural Design

In PVC structural design, balance and ballast can be used to ensure that a structure is stable and able to withstand external loads and forces.

Balance refers to the distribution of weight within the structure. A structure that is balanced will be less likely to tip over or deform under load. To achieve balance, the weight of the structure must be evenly distributed across its length and width.

Ballast is a material, usually sand or water, that is used to adjust the balance of a structure. It is typically placed in Tanks or compartments within the structure, and can be moved or added to as needed to achieve the desired balance.

For example, a PVC structure that is being used as a floating platform may have ballast tanks that can be filled with water to adjust its balance. If the structure is carrying a heavy load, it may need more ballast to maintain its balance. Conversely, if the structure is carrying a lighter load, it may need less ballast.

Airspace Model Elevation Conversion Scale

Class / Level	Airspace Class feet per Level	PVC Inch per 300 Feet	Structural Level Height & Pex Airspace Diameter
E Starts @ 700 ft AGL	700 feet	1"	2.25"H x 18" Dia.
D Surface to 2,500 ft MSL	2,500 feet	1"	8"H x 18" Dia.
C Surface to 1,200 ft MSL	1,200 feet	1"	3.5"H x 18" Dia.
C 1,200 MSL to 4,000 ft MSL	1,800 feet	1"	5.5"H x 24" Dia.
B Surface to 1,200 ft MSL	1,200 feet	1"	3.5"H x 18" Dia.
B 1,200 MSL to 4,000 ft MSL	1,800 feet	1"	5.5"H x 24" Dia.
B 4,000 MSL to 10,000 ft MSL	6,000 feet	1"	19.5"H x 30" Dia.

MODULE V QUIZ

1. **What does balance refer to?**
   A) Distribution of water within the structure
   B) Distribution of weight within the structure
   C) Distribution of the walls of the structure

2. **In PVC structural design, what can be used to ensure that a structure is stable?**
   A) Balance and ballast
   B) Efficiency and safety
   C) Flexibility and collaboration

3. **What is ballast?**
   A) A bed that is used to adjust the balance of a structure
   B) Something heavy
   C) A PVC that is used to adjust the balance of a structure

4. **What is the main goal of an airspace classification model?**
   A) To ensure the safety of all aircraft
   B) To ensure the efficiency of all aircraft
   C) To ensure the flexibility of all aircraft

5. **What is the airspace level for Class E?**
   A) Surface to 1,200 ft MSL
   B) Surface to 2,500 ft AGL
   C) Starting at 700 ft AGL

MODULE VI

AIRSPACE MODEL BUILD LAB

Tool List

Tool Description	Used For:
25-foot retractable Tape Measurer	Measuring PVC pipe
48" Paper Tape Measurer	Measuring Pex Tubing
Pressure Grip Pliers	Removing PVC Fittings
Pipe Level (clamp on type)	Verify level connections
PVC Pipe Cutter	Trimming PVC Pipes
Culinary Scissors	Cutting Pex Tubing
Fine Sand Paper	Smoothing cut edges of PVC Pipe

AIRSPACE CLASSIFICATION BUILD GUIDELINES
6FT X 6FT Model

Airspace E 6ft X 6ft Model Construction Goals

Challenge: Build a **Class E** Airspace Veri-port model using ½" diameter PVC pipe and fittings. After building the airspace model, you must fly a 90mm platform drone from takeoff position out of the airspace model following rules for airport takeoff flight path rules. You must document you model build with pictures and record your flight.

Model Criteria Requirements: Model Base 2.25"ht x 11" width with an upper 5-mile Radius red Pex ring must have an approximate diameter pf 21" inches.

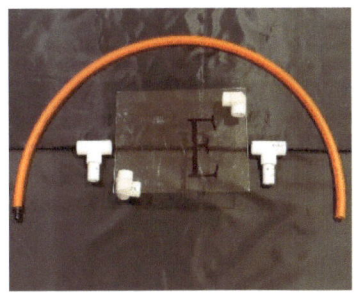

CLASS E

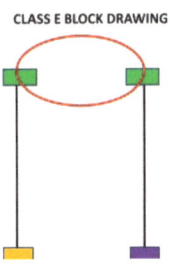

Class / Level	Airspace Class feet per Level	PVC Inch per 300 Feet	Structural Level Height
E Starting at 700 ft AGL	700 feet	1"	2.25"

Class E Parts List	Measurements	Quantity	Part #
Class E Horizonal Red Pex Tubing	½" Diameter X 42" Length	1	None
Class E Level 1 Vertical Side Pipes	½" Diameter X 2.25" Length	2	82 -83
Class E Level 1 PVC H90T	½" Diameter	2	None
Class E Base Level PVC 4P1V	½" Diameter	1	None
Class E Base Level PVC 5P1V	½" Diameter	1	None

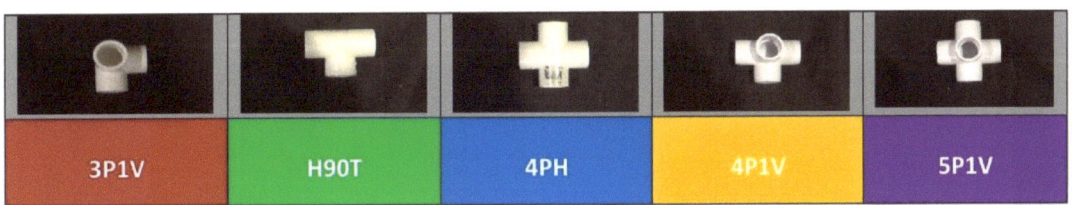

10 B

Airspace D 6ft X 6ft Model Construction Goals

Challenge Class D: Build a Airspace Verti-port model using ½" diameter PVC pipe and fittings. After building the airspace model, you must fly a 90mm platform drone from takeoff position out of the airspace model following rules for airport takeoff flight path rules. You must document you model build with pictures and record your flight.

Model Criteria Requirements: Model Base 8.25"ht x 11" width with an upper 5-mile Radius Blue Pex ring must have an approximate diameter of 12" inches.

CLASS D BLOCK DRAWING

 CLASS D

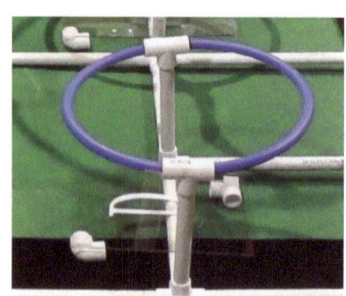

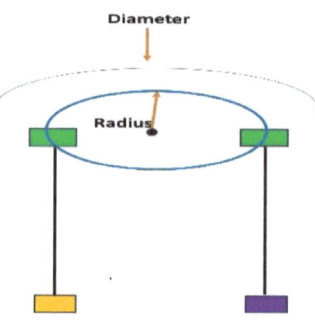

Class / Level	Airspace Class feet per Level	PVC Inch per 300 Feet	Structural Level Height
D Surface to 2,500 ft MSL	2,500 feet	1"	8"H

Class D Parts List	Measurements	Quantity	Part #
Class D Horizonal Blue Pex Tubing	½" Diameter x 42" Length	1	None
Class D Surface Vertical Side Pipes	½" Diameter x 8" Length	2	80-81
Class D Level 1 PVC H90T	½" Diameter	2	None
Class D Base Level PVC 5P1V	½" Diameter	1	None
Class D Base Level PVC 3P1V	½" Diameter	1	None

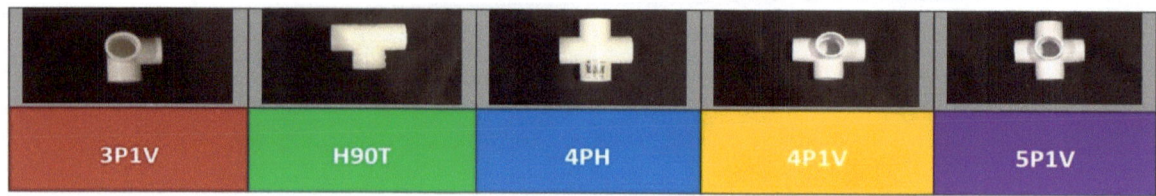

Airspace C 6ft X 6ft Model Construction Goals

Challenge: Build a **Class C Airspace** Verti-port model using ½" diameter PVC pipe and fittings. After building the airspace model, you must fly a 90mm platform drone from takeoff position out of the airspace model following rules for airport takeoff flight path rules. You must document you model build with pictures and record your flight.

Model Criteria Requirements: Model Base 3.5"ht x 11" width with a 5-mile Radius red Pex ring must have an approximate diameter of 12" inches. The second Level must extend outward 2.5" on each side horizontally and then extend vertically 5.5" to the upper Pex ring which must have an approximate diameter of 24".

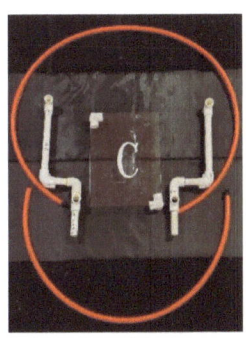
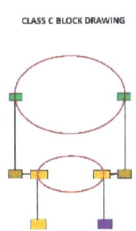

CLASS C

Class / Level	Airspace Class feet per Level	PVC Inch per 300 Feet	Structural Level Height
C	Surface to 1,200 ft MSL	1"	3.5"H
C	1,200 ft MSL to 4,000 ft MSL	1"	5.5"H

Class C Parts List	Measurements	Quantity	Part #
Class C Horizonal Blue Pex Tubing Level 1	½" Diameter x 42" Length	1	None
Class C Horizonal Blue Pex Tubing Level 2	½" Diameter x 52.5" Length	1	None
Class C Level 1 Vertical Side Pipes	½" Diameter x 3.5" Length	2	74-75
Class C Level 2 Vertical Side Pipes	1/2" Diameter x 5.5" length	2	76-77
Class C Level 1 PVC 4P1V	½" Diameter	2	None
Class C Level 1 PVC 90L	½" Diameter	2	
Class C Level 2 PVC H90T	½" Diameter	2	None
Class C Base Level PVC 5P1V	½" Diameter	1	None
Class C Base Level PVC 4P1V	½" Diameter	1	None

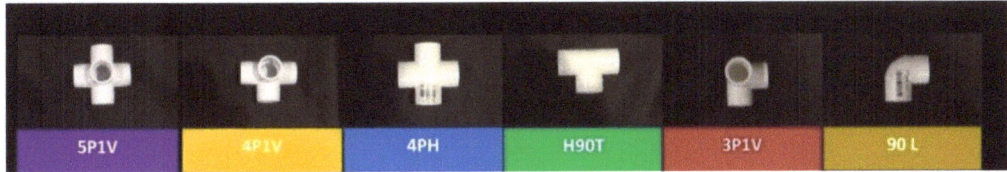

Airspace B 6ft X 6ft Model construction Goals

Challenge Class B: Build an Airspace model using ½" diameter PVC pipe fittings and ½" Pex tubing & fittings. After Building the airspace model, you must fly a 90mm platform drone from takeoff position out of the airspace model following rules for airport takeoff flight path rules. You must document you model build with pictures and record your flight.

Model Criteria Requirements: Model Base 4" height x 11" width with a first level tier Pex ring which has a 5-mile Radius with an approximately 12" Diameter. The second Level must extend outward 2.5" on each side horizontally and then extend vertically 7.25" to the upper blue Pex ring which must have an approximate diameter of 24". The 3rd tier extends upward 19.5" from the second tier and extend outward 2.5" on each side with a 48" diameter blue Pex ring.

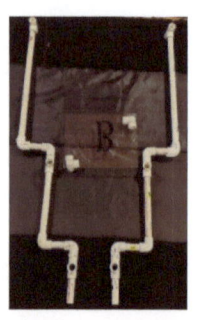

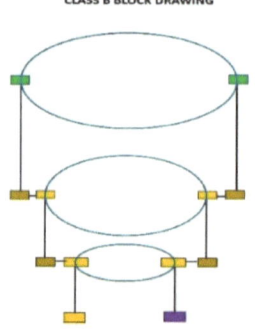

CLASS B BLOCK DRAWING

CLASS B

Class / Level	Airspace Class feet per Level	PVC Inch per 300 Feet	Structural Level Height
B	Surface to 1,200 ft MSL	1"	3.5"H
B	1,200 ft MSL to 4,000 ft MSL	1"	5.5"H
B	4,000 ft MSL to 10,000 MSL	1"	19.5"H
Class B Parts List	**Measurements**	**Quantity**	**Part #**
Class B Horizonal Blue Pex Tubing Level 1	½" Diameter x 42" Length	1	None
Class B Horizonal Blue Pex Tubing Level 2	½" Diameter x 52.5" Length	1	None
Class B Horizonal Blue Pex Tubing Level 3	½" Diameter x 63" Length	1	None
Class B Level 1 Vertical Side Pipes	½" Diameter x 3.5" Length	2	74-75
Class B Level 2 Vertical Side Pipes	½" Diameter x 5.5" Length	2	76-77
Class B Level 3 Vertical Side Pipes	½" Diameter x 19.5" Length	2	78-79
Class B Base Level PVC 4P1V	½" Diameter	1	None
Class B Base Level PVC 5P1V	½" Diameter	1	None
Class B Level 1 PVC 4P1V	½" Diameter	2	None
Class B Level 1 PVC 90L	½" Diameter	2	None
Class B Level 2 PVC 4P1V	½" Diameter	2	None
Class B Level 2 PVC 90L	½" Diameter	2	None
Class B Level 3 PVC H90T	½" Diameter	2	None

DESKTOP AIRSPACE CLASSIFICATION BUILD GUIDELINES
2FT X 3FT Model
CLASS E DESKTOP

Model Criteria Requirements: Model Base 2.25"ht x 11" width with an upper 5-mile Radius red Pex ring must have an approximate diameter of 6" inches.

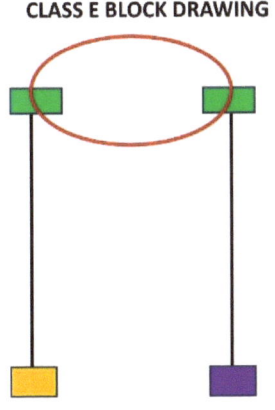

Class / Level	Airspace Class feet per Level	PVC Inch per 300 Feet	Structural Level Height
E Starting at 700 ft AGL	700 feet	1"	2.25"H

Class E Parts List	Measurements	Quantity	Part #
Class E Horizonal Red Pex Tubing	½" Diameter X 42" Length	1	None
Class E Level 1 Vertical Side Pipes	½" Diameter X 2.25" Length	2	82-83
Class E Level 1 PVC H90T	½" Diameter	2	None
Class E Base Level PVC 4P1V	½" Diameter	1	None
Class E Base Level PVC 5P1V	½" Diameter	1	None

CLASS D DESKTOP

Model Criteria Requirements: Model Base 8.25"ht x 11" width with an upper 5-mile Radius Blue Pex ring must have an approximate diameter of 6" inches.

CLASS D BLOCK DRAWING

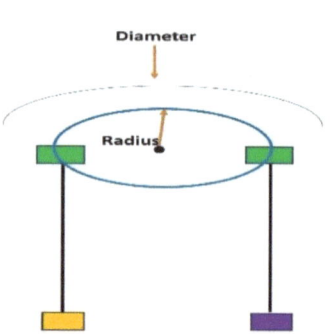
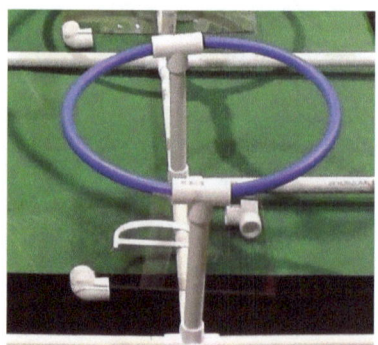

Class / Level	Airspace Class feet per Level	PVC Inch per 300 Feet	Structural Level Height
D Surface to 2,500 ft MSL	2,500 feet	1"	8"H

Class D Parts List	Measurements	Quantity	Part #
Class D Horizonal Blue Pex Tubing	½" Diameter x 42" Length	1	None
Class D Surface Vertical Side Pipes	½" Diameter x 8" Length	2	None
Class D Level 1 PVC H90T	½" Diameter	2	80-81
Class D Base Level PVC 5P1V	½" Diameter	1	None
Class D Base Level PVC 3P1V	½" Diameter	1	None

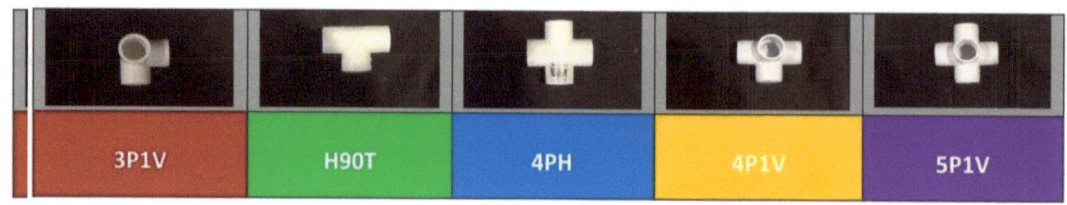

CLASS C DESKTOP

Model Criteria Requirements: Model Base 3.5"ht x 11" width with a 5-mile Radius red Pex ring must have an approximate diameter of 6" inches. The second Level must extend outward 2.5" on each side horizontally and then extend vertically 5.5" to the upper Pex ring which must have an approximate diameter of 12".

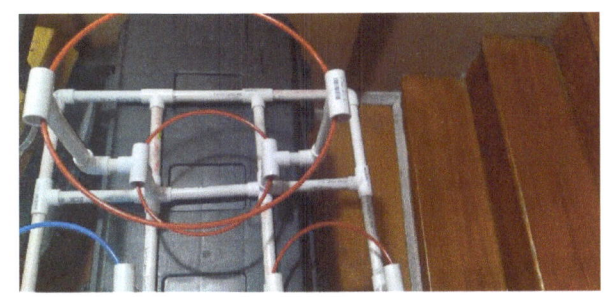
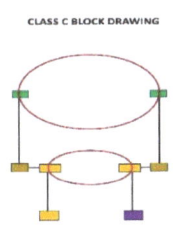

CLASS C BLOCK DRAWING

Class / Level	Airspace Class feet per Level	PVC Inch per 300 Feet	Structural Level Height
C	Surface to 1,200 ft MSL	1"	3.5"H
C	1,200 ft MSL to 4,000 ft MSL	1"	5.5"H
Class C Parts List	**Measurements**	**Quantity**	**Part #**
Class C Horizonal Blue Pex Tubing Level 1	½" Diameter x 42" Length	1	None
Class C Horizonal Blue Pex Tubing Level 2	½" Diameter x 52.5" Length	1	None
Class C Level 1 Vertical Side Pipes	½" Diameter x 3.5" Length	2	74-75
Class C Level 2 Vertical Side Pipes	1/2" Diameter x 5.5" length	2	76-77
Class C Level 1 PVC 4P1V	½" Diameter	2	None
Class C Level 1 PVC 90L	½" Diameter	2	
Class C Level 2 PVC H90T	½" Diameter	2	None
Class C Base Level PVC 5P1V	½" Diameter	1	None
Class C Base Level PVC 4P1V	½" Diameter	1	None
1 ½" PVC Nipple	½" Diameter	2	None

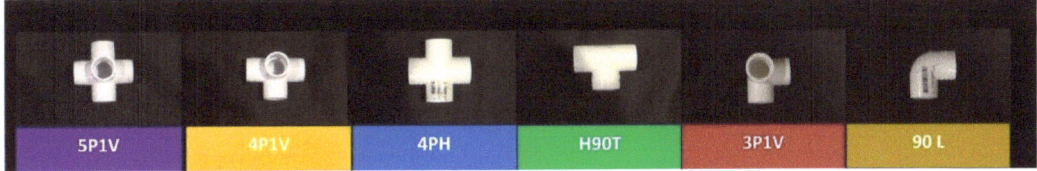

CLASS B DESKTOP

Model Criteria Requirements: Model Base 4" height x 11" width with a first level tier Pex ring which has a 5-mile Radius with an approximately 6" Diameter. The second Level must extend outward 2.5" on each side horizontally and then extend vertically 7.25" to the upper blue Pex ring which must have an approximate diameter of 12". The 3rd tier extends upward 19.5" from the second tier and extend outward 2.5" on each side with a 24" diameter blue Pex ring.

Class / Level	Airspace Class feet per Level	PVC Inch per 300 Feet	Structural Level Height
B	Surface to 1,200 ft MSL	1"	3.5"H
B	1,200 ft MSL to 4,000 ft MSL	1"	5.5"H
B	4,000 ft MSL to 10,000 MSL	1"	19.5"H

CLASS B BLOCK DRAWING

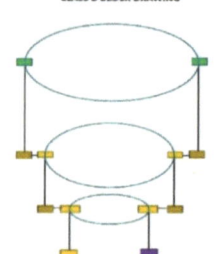

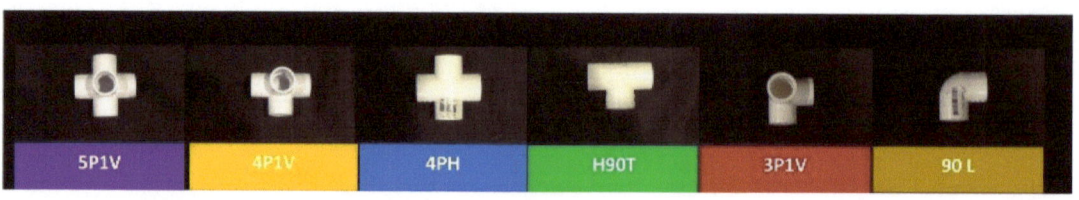

5P1V	4P1V	4PH	H90T	3P1V	90 L

Class B Parts List	Measurements	Quantity	Part #
Class B Horizonal Blue Pex Tubing Level 1	½" Diameter x 42" Length	1	None
Class B Horizonal Blue Pex Tubing Level 2	½" Diameter x 52.5" Length	1	None
Class B Horizonal Blue Pex Tubing Level 3	½" Diameter x 63" Length	1	None
Class B Level 1 Vertical Side Pipes	½" Diameter x 3.5" Length	2	74-75
Class B Level 2 Vertical Side Pipes	½" Diameter x 5.5" Length	2	76-77
Class B Level 3 Vertical Side Pipes	½" Diameter x 19.5" Length	2	78-79
Class B Base Level PVC 4P1V	½" Diameter	1	None
Class B Base Level PVC 5P1V	½" Diameter	1	None
Class B Level 1 PVC 4P1V	½" Diameter	2	None
Class B Level 1 PVC 90L	½" Diameter	2	None
Class B Level 2 PVC 4P1V	½" Diameter	2	None
Class B Level 2 PVC 90L	½" Diameter	2	None
Class B Level 3 PVC H90T	½" Diameter	2	None
1 ½" PVC Nipple	½" Diameter	4	None

Module III Airspace Build Lab Challenge Questions

1) What is the total elevation of the class E Verti-port?

a) 700 feet AGL (Above Ground Level)
b) 1,200 feet AGL (Above Ground Level)
c) 4,000 feet AGL (Above Ground Level)

2) What is the radius in miles of the surface area up to 2,500 feet MSL (Mean Sea Level) of a class D Verti-Port?

a) 1 mile
b) 3 miles
c) 5 miles

3) How many cubic feet does a 5ft ht. x 6ft L x 6ft W Navigation Airspace model contain?

a) 180 cubic feet
b) 36 cubic feet
c) 216 cubic feet

4) How much does ¼ + ½ + ¾ +1 ½ =?

a) 3
b) 2 ½
c) 3 ¾

Bonus Question

How many statute square miles does a navigation Sectional Chart contain which covers 2° degrees Latitude North by 2° degrees longitude West?

a) 10,560 square miles
b) 21,120 square miles
c) 14,400 square miles

NAVIGATION SURFACE GRID CONNECTOR PARTS INVENTORY

5P1V	4P1V	4PH	H90T	3P1V	90 L
4	4	1	13	5	0

NAVIGATION SURFACE GRID PVC PIPE PARTS INVENTORY

PIPE SIZE	QUANTITY
11" PIPE	23
23" PIPE	15

AIRSPACE WALLS CONNECTOR PARTS INVENTORY

5P1V	4P1V	4PH	H90T	3P1V	90 L
0	8	4	2	4	0

www.ingramcontent.com/pod-product-compliance
Lightning Source LLC
Chambersburg PA
CBHW040907020526
44114CB00038B/87